Employment and the Transfer of Technology

Rudolf Henn
Lothar Späth
Hermann Lübbe
Gerhard Krüger

Employment and the Transfer of Technology

Springer-Verlag
Berlin Heidelberg New York Tokyo

Prof. Dr. Rudolf Henn, Universität Karlsruhe
Ministerpräsident Dr. h.c. Lothar Späth, Stuttgart
Prof. Dr. Hermann Lübbe, Universität Zürich
Prof. Dr. Gerhard Krüger, Universität Karlsruhe

The German edition was published under the title
"Beschäftigung und Technologietransfer"
© 1985 Athenäum Verlag GmbH, Königstein/Ts.

ISBN-13: 978-3-642-71294-4   e-ISBN-13: 978-3-642-71292-0
DOI: 10.1007/ 978-3-642-71292-0

Typesetting: K. Triltsch GmbH, 8700 Würzburg

2142/3140-543210

Contents

Commemorative Address on the Occasion of the Presentation of an Honorary Doctorate to the Minister President of the State of Baden-Württemberg, Mr. Lothar Späth, on June 22, 1984

RUDOLF HENN

The Faculty of Economics at the University of Karlsruhe has decided to award the Honorary Doctorate of Economics to Mr. Lothar Späth. As the Dean of this Faculty, I therefore have the honor today to open the ceremonial conference and to deliver the address.

When someone asks why the Faculty of Economics at the University of Karlsruhe is presenting an honorary doctoral degree to the Minister President, it must be said first of all that Lothar Späth's conceptions as well as realizations of economic models have influenced a number of papers and theses which have been developed at this faculty, whose members include not only economists and managers but also computer scientists, mathematicians, and engineers. The exchange of ideas between our faculty and a politician who sees the solution of today's pressing economic problems to lie in connection with new technological developments has been fruitful indeed.

As an economic politician, Lothar Späth evidently asked himself the same question which was considered before him and is considered by many federal economists in our days as being the central problem of the economy, namely, how to promote economic growth, achieving full employment, throughout the course of the business cycle.

Let us look at Späth's reflections in association with former and present economic theories.

The study of economic history shows that as early as the beginning of the nineteenth century, structural crises take place in connection with industrialization. These crises are

caused by the fast change of production techniques and location possibilities due to the appearance of new means of
transportation. Not only do they differ from the economic
crises of the preindustrial times, which were mostly caused
by natural catastrophes[1], but they often demand specific economic and political instruments for their circumvention.
There is, however, no uniform business cycle theory, similar
to equilibrium theory, but a multitude of business cycle
theories.

An external feature which could already be seen during
previous structural crises is overproduction in the traditional
branches of the economy causing stockpiling of mass quantities.

David Ricardo considered this overproduction to be a temporary and partial interruption of the equilibrium and thus
of no particular theoretical interest.[2] According to Thomas
Robert Malthus, however, the crisis of general overproduction develops as a result of an excessive accumulation of
capital which itself is caused by an uneven distribution of income. According to Malthus, this produces a decline in the
demand-side behind the supply-side of goods. In other
words, the corollary of a theory of overproduction is a theory
of "underconsumption."

Malthus' ideas were adapted and used by Marx to postulate the breakdown of the economic system. In so doing
Marx extrapolated the economic structure of his time without
himself presenting any starting points for a solution.

The Swedish economist Knut Wicksell supplements the
overinvestment theories of classical economists, such as
Hayek, Machlup, and Röpke, with the inclusion of monetary
variables. Central is the distinction between the money market interest rate which is fixed by the banking system, and
the "natural" interest rate which acts as an equilibrium price,
thus equalizing the supply of savings and the demand for in-

---

1 See K.E. Born, Wirtschaftskrisen. In: Handwörterbuch der Wirtschaftswissenschaft, 1982, vol. 9, pp. 130–141
2 See G. Stavenhagen: Geschichte der Wirtschaftstheorie, 1964, 3d edn.,
p. 508

vestment capital.[3] A drop in the money market interest rate below the level of the "natural" interest rate results in an overproportional expansion of capital goods in relation to consumer goods. The oversupply in the investment goods sector can not be absorbed by the consumer goods sector. According to Wicksell, a new equilibrium can only be restored by a correction of the distorted production structure, namely, by an increase of the money market interest rate up to the level of the "natural" interest rate[4]; this reduces corporate investments.

On the basis of empirical investigations, Keynes did not agree with Wicksell's conclusions that, following the collapse of investment activities, the renewed decrease of the money market interest rate stimulates the profit expectations of investors enough to start a new recovery. Keynes, therefore, suggests utilizing interest-free independent public investments to stimulate employment and thereby consumption. This idea can be explained by means of a simple diagram (Fig. 1).

---

3 See H.-J. Vosgerau, Konjunkturtheorie: In: Handwörterbuch der Wirtschaftswissenschaft, 1978, vol. 4, p. 407
4 Ibid.

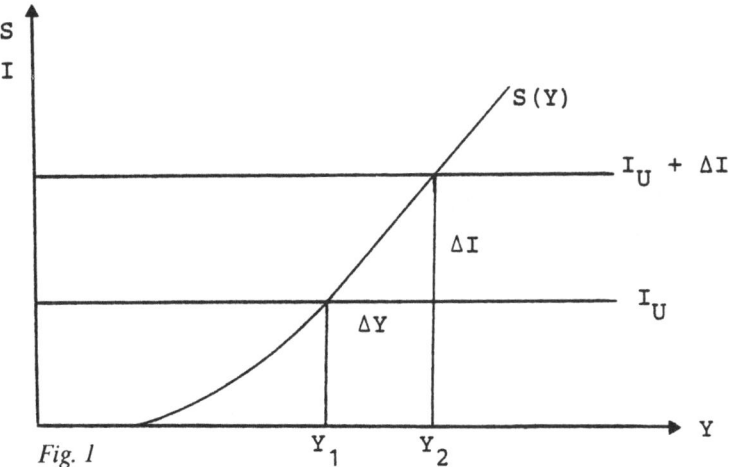

Fig. 1

Let I_U represent corporate investments subject to interest rates which, when undertaken at a given savings function, lead to an equally balanced national income Y_1. Autonomous, interest-free governmental investments ΔI result in an increase of income

$$\Delta Y = \frac{1}{\dfrac{\Delta S}{\Delta Y}} \Delta I$$

and thereby at the same time result in an increase of employment and consumption.

The quotient $\dfrac{1}{\dfrac{\Delta S}{\Delta Y}}$ is the well-known "investment multiplier" of Keynes. Keynes' conception did not remain a theoretical model. It proved successful in circumventing the world economic crisis and determined the economic and political policies of the industrial nations for over 30 years. Even as late as 1967, Keynes' idea of adequate economic growth made possible by comprehensive state control was recognized as the law for stability.

Paul Samuelson supplemented the multiplier model with a dynamic analysis of the principle of acceleration according to which corporations at all times want to adapt the amount of their capital stock to the size of the increase of consumption.[5] When at the beginning of the 1970s it turned out that economic reality could no longer be regulated by Keynes' instrument for comprehensive control, economic policy came back to Friedman's concept of fighting inflation with the acceptance of a natural unemployment rate. Coupled with a steady increase of the money supply, this policy indeed succeeded in opposing the inflationary development which could be observed internationally.

---

5 See P.A. Samuelson, Interactions between the multiplier analysis and the acceleration principle, Review of Economic Statistics, 1939, and E. Schneider, Einführung in die Wirtschaftstheorie, vol. 3, 12th edn., 1973, p. 238

In the face of a rapid and ever-increasing unemployment rate, however, the thesis of natural unemployment proves untenable.

Without involving the well-known discussion of the Philips curve, it can be said that economic politics in Friedman's terms have shown that an inflationary monetary policy does not lead to a reduction of unemployment. Nevertheless, it is possible to acquire stability of the value of money at the cost of employment (Fig. 2).

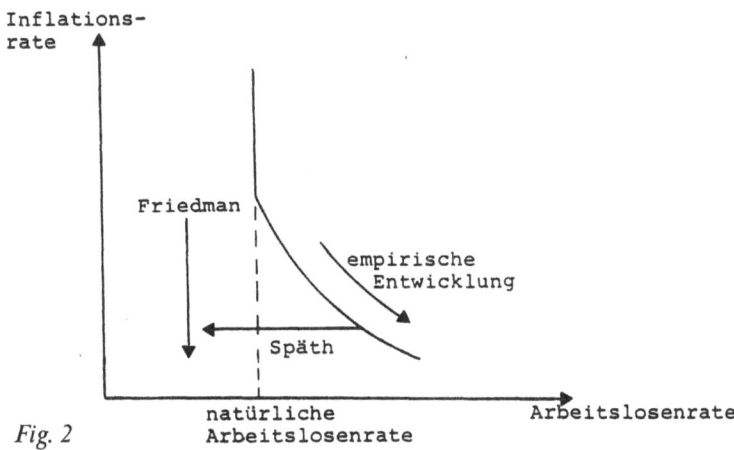

Fig. 2

The vertical arrow in Fig. 2 reflects Friedman's concept; a movement along the Philips curve in the direction of an increasing unemployment rate can be empirically observed.

As Minister President, Lothar Späth faced the problem of reducing structural unemployment without affecting the stability of the value of money – indicated in Fig. 2 by the horizontal arrow to the left. Mr. Späth recognized early the structural causes of recession and unemployment and thereby the necessity of a structural change. In his concept regarding a solution, the interconnection between technology and economics plays a decisive role. Späth sees the chance for a restructuring of the economy and for its adaptation to the fast-changing environmental conditions to lie in a transfer of technology to secure a competitive advantage for the long run. This means a strategy of directly and smoothly imple-

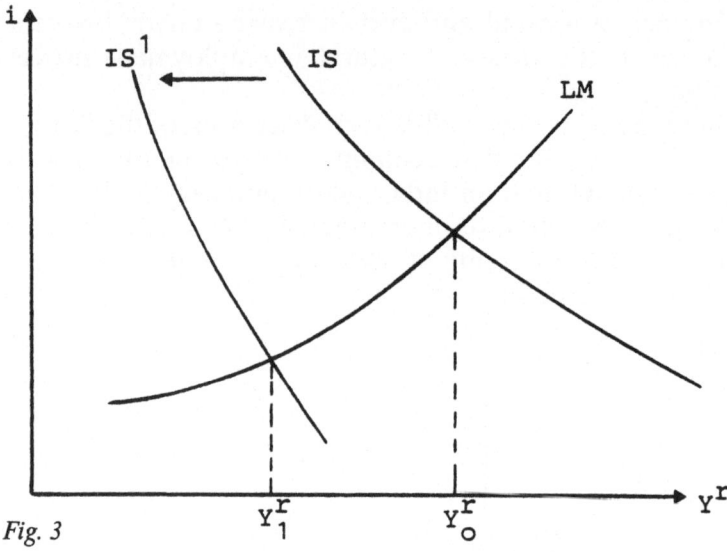

Fig. 3

menting scientific results up to the development of new industries which could be competitive in the international market and, because of their innovative character, lead to an overdemand for skilled workers.

Späth's model concerns promotion of the process of adaptation, or in other words, specific future-oriented supply-side support. In this case the government has the duty to provide a starting impulse for new technology through the creation of an appropriate infrastructure. The basis for this idea is the fact that, since the end of the full-employment phase, there has been an unwillingness on the part of corporations to investment in almost all sectors of the economy, which can not only be explained by the high capital interest rates but is also an expression of ambiguous future expectations.

Transferred to the illustration of Hicks' standard diagram (Fig. 3), this causes a shift of the IS curve to the left together with a noticeably low interest rate elasticity in the neighborhood of the equilibrium.[6]

---

6 The standard diagram by Hicks describes the equilibrium of the goods and money markets. The IS curve is the set of all pairs of interest rates i and real national income Y^r where savings S are equal to investments I. Accordingly the LM curve describes the set of all pairs (i, Y^r) where money supply M is equal to money demand L

In such a situation of uncertainty the monetaristic set of instruments does not allow any help. The old counter-measures of Keynes against unemployment and stagnated sales markets, namely, to bridge the gap of supply by public-sector investments, also proves to be useless since in this case the willingness of the entrepreneurs to invest can not be reobtained. On the contrary, the process of crowding-out, i.e., the repression of private by public activities, increases.

Today we live in a period where large, prominent German companies shy away from risky new investments and prefer to realize easy and secure profits from investments on the capital market. Therefore it is not astonishing that young researchers in industries with a potentially good future, like microelectronics, information technology, or gene technology, face formidable problems in implementing their ideas, from conceptualization to the time where the product is ready for market. To overcome these problems they require financial support, so that infrastructural facilities and production necessities are guaranteed. The government should provide financing through companies established for the depreciation of risk capital and the financial participation of state-owned credit banks. To date this model has been implemented in its initial stages, with the creation of technology centers.

Using economic terminology, it can be said that we are now at a point on the production function of the whole economy[7] where the foremost requirement is a clear, innovative improvement of the quality of the capital stock allowing the process of economic expansion to be possible on a higher level in the future. Nevertheless, this can not be accomplished by any single enterprise by itself.

However, up to now, only the impetus has been described but not the process development in terms of a model for the whole economy. Two additional conditions have to be

---

7 The partial production function for the whole economy for the production of real income Y^r given a certain amount of labor $\bar{N}$ depends on the capital stock K. The latter can be increased by real investments like the ones represented by public infrastructural output I_δ^r, i.e., in the form of technology factories

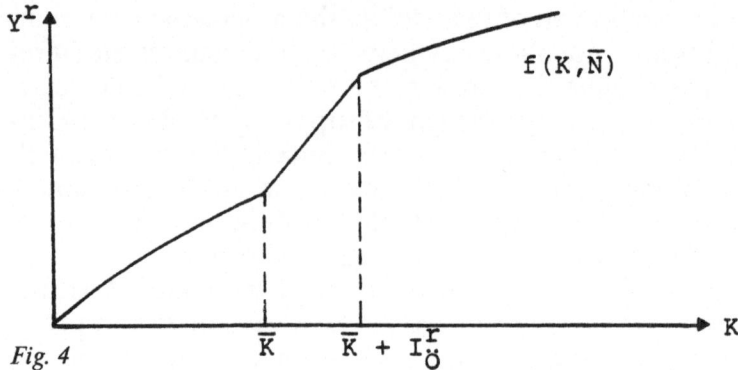

Fig. 4

guaranteed so that once made the initial start can lead to lasting economic growth:

1. On a higher capital level where $K_1 = \bar{K} + I_\delta^r$, the physical marginal productivity of labor N is greater than on the original level $\bar{K}$, i.e.,

$$\left.\frac{\partial f}{\partial N}\right|_{(K_1, N)} > \left.\frac{\partial f}{\partial N}\right|_{(\bar{K}, N)}$$

2. The expansion of the economy which leads from the IS¹ curve to the higher level – represented by the IS² curve – is not caused by curtailing production. Moreover, the equally weighted sales increase ΔY^r (caused by the public support of I_δ^r) calls for additional labor input ΔN (see Fig. 5).

Although condition (1) can be considered the less problematic, condition (2) has to be examined more carefully. It is immediately evident that for those industries in the economy which did not previously exist, the effect of employment curtailment will not prevail with the promotion of I_δ^r.

Since there were no places of employment, they could not have been economized. On the contrary, these places of employment have to be newly filled. Additionally, new products normally provide for new markets, and demand – consider the case of communications technology – which previously did not exist is satisfied. Even in traditional industries, e.g., mechanical engineering, new markets can be opened through changed requirements stemming, for example, from the pol-

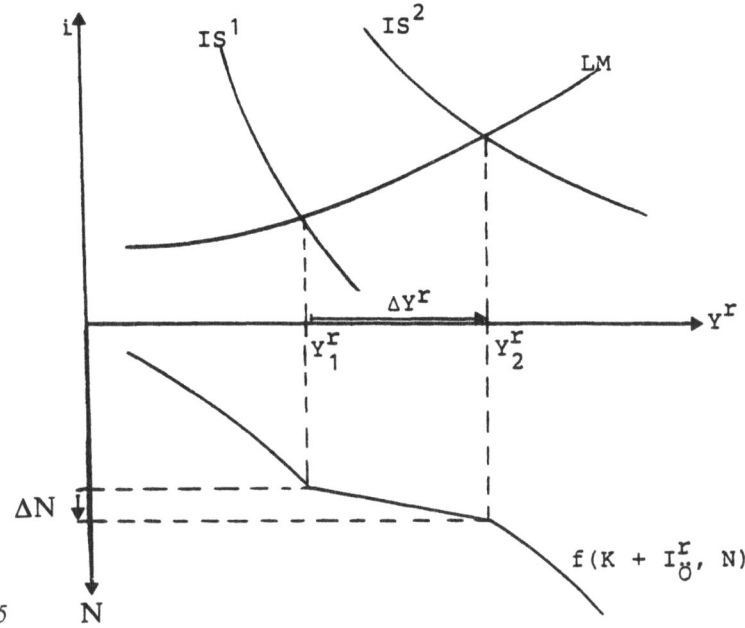

Fig. 5

lution control area. In other words, an economic policy in support of innovative production areas, started by the aforementioned impulses, during its development will not only generate a higher GNP but also more jobs, thus integrating more workers.

The further development of the economic events describing the expansion starting at S_2 can be interpreted as follows (see Fig. 6).

The innovative public investments I_δ^r induce additional innovative investments of private enterprises I_U^r starting at the point in time where the capital supply $K_1 = \bar{K} + I_\delta^r$ and the national income $Y_2^r = f(K_1, N)$ are simultaneously available. The investments will be promoted as long as the GNP value is higher than the production costs, i.e.,

$$p \cdot f(K_1 + I_U^r(t), N(t)) - p_n \cdot N(t) - p_K \cdot (K_1 + I_u^r(t)) > 0$$

$$= > \frac{dI_U^r}{dt} > 0$$

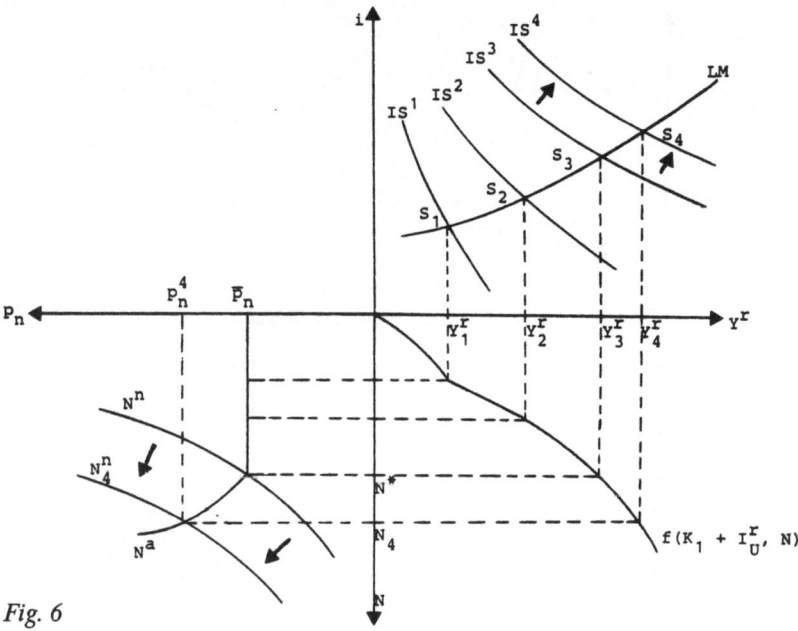

Fig. 6

Hereby we signify p as the price level for goods, p_n as the nominal wage rate, and p_K as the price per capital unit.

The labor market can be described by a labor demand function N^n of corporations which is oriented towards real wages $(N^n = N^n (p_n/p))$ and by a labor supply function of the workers which is oriented towards nominal wages $(N^a = N^a (p_n))$. The labor supply function is fixed until N^*; it increases in a strictly linear function for greater N. Innovative investments mean a further shift to the right of the IS curve to IS^3 and by this an expansion of the production to Y_3^r at constant nominal wages. Workers will allow further expansion of production only if wages increase. The labor demand function of the corporations, caused by the productivity increases due to innovative investments of the corporations, shifts outward. Further expansion of production can go on until the point S_4 where higher wages are equal to the value of the marginal product of labor. This relation determines the demand N_4^n. The situation marked by S_4 represents the final equilibrium of the money, goods, and employment markets under full employment.

Summarized, one can say that the infrastructure made available for the development of technology makes expansive economic development possible and improves the employment situation decisively. However, in the end, it is limited by a ceiling caused by the later demands for wage increases.

In spite of a plethora of publications in recent years on the question of structural change and unemployment, no other author has seen the economic and political implications to be drawn from the present situation with such clarity as Lothar Späth.

Späth's model connects elements of a modified multiplier theory together with Friedman's concept of supply-side support.

The creation and promotion of newly organized structures and economic conditions for the cooperation between research and economics has earned Mr. Späth the credit he deserves from science, and especially from our faculty. His personal effort in these areas can be seen clearly from multiple publications and lectures, but also from the organization of workshops and conventions, e.g., the "Conventions of the state government of Baden-Württemberg concerning future prospects for an industrialized country" in 1982 and 1983. In his opening lecture for the first of these he said: "We have to accept the technological challenge which requires concentrated cooperation between government, economics, science, and society."

Späth views the responsibility of politics in a very broad sense, in terms of time and dimension. Guaranteeing future opportunities includes the creation and long-term support of foreign markets – through the export of education in its various forms be it foreigners studying in Germany, the export of technology, or the exchange visits of scientists at international research institutions.

The Faculty of Economics honors Mr. Lothar Späth through this presentation of the honorary doctor of economics for his achievement in the development and enforcement of new economic and political concepts and for the development of new possibilities for the cooperation of science and economics in future-oriented politics.

Acknowledgement of Commemorative Address

LOTHAR SPÄTH

Magnifico, Dear Professor Henn, Ladies and Gentlemen!

I would like to thank the Faculty of Economics at the University of Karlsruhe for the high honor it has granted me today. I do not want to conceal the fact that I had hesitated quite a bit at first to accept this honorary doctorate, although I was very pleased and honored by the decision of the faculty. My hesitation was mainly caused by my position as a politician who faces science more often as a person asking questions and looking for solutions than one offering completed concepts and finalized solutions. On the other side – and this finally overcame my doubts – for me, the permanent dialogue between science and politics is one of the presuppositions of a well-functioning modern society. Looking at the highly complex problems which we face, this dialogue cannot be close enough – without affecting the freedom of the one and the decision-making responsibilities of the other.

Allow me to relate some "straight talk" about the topics which presently play a major role in economic and financial politics and as mentioned in your address, Professor Henn. The question of in which direction and with what means the government could or should influence the administration of economic and technical structural change is presently being discussed in a strong and quite controversial manner. There are not only the divergent economic schools with their different emphases on the supply and demand sides, but in the area of regulation politics there are also lively arguments between advocates of a policy which should rely strictly on governmental subsidies and those who would like more support on a preventive and interventionist level.

I believe that satisfactory results can only be achieved not by creating concepts out of regional and sectoral piecemeal goals but by trying to conceive of political instruments as derived from worldwide economic, scientific, and technical development trends. The first alternative, to which I incidently object, would mean, for example, offering provisions in accordance with current regional requirements or agreeing to a policy of subsidies to preserve the current structure (in which the target is, say, securing short-term employment) – even if (due to their high costs) these industry structures do not show much chance of survival in international competition in the long run. In my opinion, the inefficiency of such measures prohibits their application within the framework of governmental subsidy politics. I also fail to recognize how far they could be accompanied by a higher economic elan as compared with the results of so-called interventionist actions by government, conducted on the basis of a policy actively promoting science and technology.

The second alternative would support a process of economic adaptation based on international comparative standards; safeguarding a public policy of such support would in a sense affect the infrastructure. This would require substantially increased flexibility and, in comparison with a policy which just subsidizes current structures, an inclination towards risk. Strictly speaking one has to anticipate the future. This is not possible without a sufficient and clear perception of how the international distribution of production and work will develop and what kind of repercussions this will have on the competitive capability of the domestic economy. I willingly concede that errors in connection with such assessment are certainly possible, since instruments for forecasting are limited. On the other hand, the situation does not seem right to me whereby government, by trying to conduct a future-oriented politics of economics and technology (and thereby inevitably stressing certain structures), quasi-pursues a sectoral-choice policy in accordance with the perceived criteria of growth – which of course would be suspect with re-gard to the concept of a free economy.

Anyone who argues in such a way ignores the fact that in a free economy, even under the active and well-aimed pro-motion of science and technology, the state never supports particular products and their sales. Instead it merely provides infrastructural preconditions for new product development possibilities which in principal everyone can take advantage of. In effect this is also a matter of creating suitable economic conditions similar to those in the area of corporate taxation – currently demanded by many people and which are correctly classified as classical, free economy instruments for global control. The only point here is that the economic conditions of a new entrepreneur, trying to get established and whose main concern in the first few years is expansion of his pro-duction capacity and not the taxation of his often nonexistent profits, look different from those of an enterprise which has been in the market for a long time and which has to decide the question of reinvesting its capital.

The one case deals with start-up assistance, the other with ease of adaptation – and either both fit in the free economy concept or neither does.

As you can see my considerations always revolve around two terms which I indeed believe have a central economic and political meaning: the development of the world's econ-omy and the creation of new infrastructure. I believe both are more closely related than is often considered. We are on the way to a quickly integrating world market. The signifi-cance of transportation costs and the fact that location con-ditions are tied to available resources of a natural or an in-dustrial type are balanced quickly. In exchange other factors clearly become more important: wage expenses, density of communications, proximity to research and development fa-cilities outside of plants, stability and type of social order, and receptiveness of economic conditions to accommodating innovation.

In other words, entrepreneural decisions for investments are taken more and more under consideration of long-term perspectives of expansion and in weighing the international comparative advantages.

This leads to far-reaching repercussions in considering the question of what a modern domestic infrastructure should look like. The connection between transportation infrastructure and the location of industry is truly nothing new. It can be followed from the development of sea ports leading to cities of trade, to the rise of industrial complexes at railway junctions, to the present localization of industrial areas along highways and major roads. In fact it is also not new that, through the promotion of particular infrastructures, government in all of these cases pursued practical economic politics at the same time, without anyone arriving at the idea of undue manipulation of the market, or to use an earlier phrase, of the "flow of goods."

On the contrary, the economic expansion of the market regularly followed the preceding creation of new infrastructures which could not have been realized by individual corporations. What is possibly new under the aspect of a world market in a state of integration is the quality of these infrastructural changes. The field of communications and the rapid availability of the potentially extreme consequences of science and information are playing ever-increasing roles. In this case the question of what technical standards of communications a society has and how the connection between economics and science is organized becomes immediately relevant for competition.

This is the background for the strong effort which we have been conducting for quite some time in the area of technology transfer and in the creation of new communications networks. Scientific analyses have repeatedly shown that neither the quality of our research nor our product quality, on the whole, have cause to shy away from comparison with Japan or the USA. Where we fall short is in terms of a "functional connection," i.e., the assignment of performance and result-oriented tasks between public, scientific, and economic sectors. For a long time the promotion of research by the government was too widely scattered. There was too little emphasis on promoting an "innovative impulse" and there was too much rule by the struggle of dogma, i.e., between those advocating direct and indirect financial interventions. If one compares research orders of the government of Japan (e.g., for the development of a new generation of computers) or the United States (e.g., in air and space technology), one quickly recognizes a much more pragmatic and systematic process.

In pure science the distance from economic reality is of course particularly great and incontestable. Nevertheless, politics in particular should be careful not to reproach science for this. The traditional self-assessments of colleges and universities as well as university and state laws favor a tendency for academic institutions to remain in an encapsulated world. The misconducted educational reforms and the more or less powerless acceptance of an irrelevant, performance-reducing ideology in the years between 1968 and 1973 made economic and practical research nearly impossible in Germany. This was just the time when at Japanese and American universities the path into the age of microelectronics was smoothed and also immediately followed by an unproblematic transposition into production structures – like Silicon Valley or Route 128 (near Boston). What it means to be 10 years behind in a technology which itself is not much older and which in this period has already caused worldwide structural changes can hardly be measured. Nevertheless there is no cause for resignation. One just has to know where one stands and what kind of opportunities present themselves. For example, there would be no sense in trying to link together universities and industrial enterprises in order to get on the "bandwagon" of mass production of computer elements with, say, one million circuits. This "wagon" is already gone and whoever would try to jump on it would risk getting seriously hurt. On the other hand, there are still many niches and gaps in connection with the development of custom-made microprocessors, system solutions, artificial intelligence for robot and sensor technology, steady state physics, and development of new materials in gene and biotechnical research, as well as in many other areas which you as representatives of the sciences can better assess than I. The government which tries to help here by means of research promotion perceives a real opportunity to stimulate innovation, new products and processes, and thereby, new jobs.

In this chain between scientific innovation and marketable products there certainly are still some weak points which could endanger or prevent economic success. It would not be at all tenable, for example, from a financial, economic, and political point of view to underwrite the research with considerable tax money and then simply leave it to chance for the innovation to actually trigger the desired economic impulse (which in turn per chance could provide relief to the employment market). A government acting in this way could be correctly reproached for wasting public resources. Therefore an integral part of the logic of a total concept – unfortunately this is often overlooked in discussions about the politics of technology – is to mesh research promotion as closely as possible with actions supporting technology transfer and the facilitation of technology-oriented structural changes. All three components belong together. Otherwise the danger would be great of merely pursuing "l'art pour l'art" and thus missing the entire hoped-for economic effect of this promotion.

From this logical chain of effects, namely, research promotion, transfer of technology, and aids for the establishment of enterprises, a certain infrastructure necessarily results. It shows that colleges and universities could be the potential loci of crystallization cores for new entrepreneurial activities in the so-called "high technology areas." Technology plants and technology parks, as developed in the USA and as are about to be developed in our country, do not originate from a particular mood or from an anglophilic impulse of imitation on the part of some politicians, but correspond with the development of a certain basic decision about research and economic policy.

The set-up of a new communication infrastructure – and hereby I already want to finish my economic and political remarks which really should be just remarks – is a similar case. Naturally a technology that makes it possible to shorten and to expand the paths of information and to totally let go of some conventional transmission and transportation systems must have far-reaching effects upon the structure of the economy overall.

We know about the technical medium, namely, the use of copper cables and later on glass fibers, and we know the direction of the technical progress which is leading towards multifunctional devices which integrate many different services. It is as if one starts to build roads because he knows that in a few years there will be generally available, inexpensive, and fast cars. It is thus obvious to ask: What effect will this have on the production of goods, transportation systems, design and organization of work-places, international competitive ability, and social and cultural living conditions in municipalities and families?

Fundamentally we are not much different from the people of the first industrial revolution. We wait and see what technical changes take place – many of us with the secret hope that they will not be too many – and then adapt under pressure of the new technical and economic facts. Strictly speaking the social sciences and humanities must daily assail politicians with scenarios and proposals for organization in order to insure that sociopolitical and cultural innovations would easily follow the recognizable new technological infrastructures without interruptions. If I perceive it correctly, very little of this has occurred, thus increasing the fear that in the next years, especially in the face of higher unemployment, political and social resistance against innovation will increase rather than decrease.

With this I'm again at the statement I made at the beginning, namely, that the permanent dialogue between science and politics is one of the main requirements of a well-functioning modern society. This dialogue must not be interrupted and has to give politicians the facility to push risky ideas on and to arrive at risky decisions. The possibility of being wrong is in the very nature of every decision. To do nothing, however, would be wrong for sure.

Therefore we must decide to accept for ourselves the path of "calculable risk." We owe this to the people for whom we work and who have transferred this responsibility to us.

Elites in an Egalitarian Society

Hermann Lübbe

The present German system of science developed substantially in the 1960s and 1970s. Today 2.7% of our GNP is spent on research and development; this figure stands at the forefront in international comparison. Thus, it can be stated that any deficiencies in our scientific system can hardly be caused by deficiencies in our willingness to allocate money for science. Rather, one has to look for the reasons in a faulty distribution of the money so-allocated.

Our university system in particular has expanded. The specification of detailed figures – from the development of the relevant public budgets to the number of new universities established, to the expansion of demographic sections of student group in the respective age groups – would be superfluous here.

The analogous development of secondary education, namely, the school system, is well known. For example, the portion of school children attending "Hauptschule" has been shrinking continuously. The understanding that completion of higher-level schooling is advantageous in the future pursuit of a career, including the corresponding higher income expectations, has spread with a surprising speed among parents and adolescents. Accordingly, the amount of one's lifetime spent at schools and universities has been increasing continuously. Correspondingly, the period of employment during one's life time has been shrinking. In short, high school attendance, the "Abitur," and academic studies have become the norm for broad social classes. There is no question that academic studies or even attendance of higher schools has an elite character.

This development, well known to all of us, was not unplanned and uncontrolled. It was intended in correspondence with internationally observable trends. There was no considerable resistance against these trends which in the meantime have of course stagnated. The legitimate power of the phrase "equal opportunity" proved to be irresistable in education and political science. Even when prudence would have recommended moderation of the speed of the convergence of values concerning decisive educational and political aims, public remarks were only gradually expressed, particularly by us Germans with our well-known talent for nonpragmatic radicalism, even during times of positive action.

We often hear complaints regarding the deficiencies of elite excellence particularly concerning the current German sciences. Is it the broad development of our school and scientific educational system that is to blame? For some time now, we have again and again been confronted with comparisons of figures in the media and in the scientific literature which turn out awkwardly for the German sciences. In the meantime, it is well known to all who read the literature that Germany comes off badly in comparison with the number of Nobel prize awards going to Great Britain or more than ever to the USA since the end of World War II. In addition, one knows that publications by German scientists, especially publications in German, are underrepresented within international citation indexes. In connection with this, the apprehension is well grounded, that in certain research areas, where particular impulses result in modern industrial and economically relevant developments, Germany in the meantime may have fallen hopelessly behind the USA or perhaps also Japan. This is especially true in research areas which are relevant for gene and information technology. Meanwhile the publicity of such pertinent problems has grown to such a degree that our politicians are forced to consider these problems in their election campaigns.

I wish to consider the following questions: Are the above-outlined trends of mass academia what is hindering our long overdue creation of elites? Does the broadness of our system of scientific education have a mediocratizing effect on our top science students? Do programs for scientific, educational, and political realization of opportunities contradict programs for the promotion of elites? Is it possible in the study and research of science that one gets only one of the two – either quantity or quality?

In principal the matter is a different one. Even for science it is also true that the probability of excellence in education and science principally is not inversely proportional to their breadth in society; on the contrary it increases. Were it different, the scientific and cultural scenery of the USA for example, which is correctly yet worriedly admired by our promoters of an elite, would have to remain inexplicable. Instead, the USA serves as an example for promoters of our elite because of its ability to produce a scientific elite. At the same time however the USA is a country whose scientific culture has possessed both sides for a long time: mass academic education on the one hand and elite colleges and universities on the other, academic degrees on a broad level and frequent Nobel-prize awards as well.

How does that fit together? There are certain areas of life where the effect of excellence achieved only by means of a broad fundamental base is evident and familiar to everyone. Sports for example is such an area. Whatever the advantages and disadvantages of mass sports may be, it obviously increases the chances of talent searchers to have success. It would also be absurd to assume that the fact that chess is an object of a mass culture in Russia decreases the chances of the Russians to hold their ground at international chess championships. Evidently it is just the other way around. In general terms this means that the quantitative expansion of the promotion of interest of a particular area does not decrease the chances of finding and promoting excellence, but increases them. Speaking somewhat more technically this means that the selectivity of a system increases with the quantity of comparable and thereby competitive elements.

It is not recognizable why this fundamental social connection between breadth and excellence on levels of comparability should not also be effective in science and culture. The sciences are especially subject to that kind of dialectics of equality which we are well acquainted with through the history of economics as well as the history of modern democracy. The social process of egalitarianism was not directly started by supporters of equality who destroyed the privileges of guilds and estates. On the contrary egalitarian supporters of a liberal public and social order triggered processes of social and cultural differentiation which reached all the way to the individual person. The principles of equality along liberal lines do not define an aspect in which all differences of talent, interest, and origin have to be cut down following a lawn-mowing principle. On the contrary it is the social function of these principles of equality to produce a level of comparability on which it can now be seen without the falsification of the privileges of guilds and estates or, in other words who we are and who we are not, what we can and can not do, for what we can or can not be moved to by our interests and characteristics of origin.

For an historic illustration of this elementary connection between the demand for equality and differentation to the creation of elites I would like to call to mind the reformers of the Prussian university system, whose activities resulted in the establishment of the Friedrich-Wilhelm-Universität in Berlin which was especially successful in the sciences. Here as well as in the philosophy of education and science of the European "period of enlightenment," the educational institutions were supposed to be an especially effective instrument in undermining the guilds and estates of the old society. Civil principles of equality were realized in which the path to better jobs was opened to the public for all people who successfully passed a state examination instead of just people who belonged to the upper classes. Intellectual excellence and flexibility in the areas of science, administration, and economics became more decisive for careers and job positions than mere titles of nobility.

In any case no matter how this was ideally planned, and irrespective of the breakdown of this ideal in reality (which we do not have to consider here) it still holds true that the social and cultural history and especially the history of science could not be conceived of without the equalizing and differentiating effect on all people placed in front of an examining commission.

To summarize, egalitarian principles set processes of differentiation free, if they are practiced. This is exactly why, in opposition to widespread prejudice, there is fundamentally no adverse relationship between equality and freedom. On the contrary these principles of equality have a liberating effect. If this is the case, the socially effective creation of equality of opportunities in the educational areas of schools and universities, the dominant main target of school and university politics in the past quarter of a century, should not have had an equalizing but a more differentiating effect and thereby should have promoted the creation of an elite. We actually should have reached two things at the same time: both a broad scientific education and the promotion of a scientific elite. Our scientific and educational system would have shone with egalitarian excellence.

Why is the shine not so bright? To word it more carefully: What is the reason that, according to the stated opinions of critics, the power of our educational and scientific system to produce an elite has remained so weak? What obstructed the quality of our educational system while it expanded broadly? Before answering this question I want to mention that the German scientific system is actually better than its reputation. It is in vogue today to complain loudly about deficiencies in research results. In contrast, well-deserved recognition is bestowed much less. Nevertheless, the promotion of an elite is missing in our country and next I would like to mention and explain some of the reasons for this situation.

First, in terms of many politicians and their parties there has been and still is much resistance to the effects of principles of equality which in the end inevitably lead to differentiation. This resistance is partly caused by a lack of understanding of the connection between equality and freedom which would have this type of differentiating effect. This is partly the result of certain traditional and ideological orientations. One way or the other they destroy a social and cultural condition which is an imperative requirement for the successful creation of an elite. The main thing is that these differentiating effects of equality principles must be socially and culturally accepted.

Lack of this acceptance has been the case so far. The conclusions can best be explained by examples. In our school system, we know about the "integrated schools" as institutions which, in the opinion of their supporters, increase the likeliness of opportunities in a very efficient way. If at the same time however one associates an expectation of the promotion of results with an improved likeliness of opportunities, then the people who professionally carry the responsibility for this promotion, namely the teachers, are trapped under the pressure of contradictory requirements. The professional work of a teacher in recognizing and promoting individual talent can not be performed if it must be done under the pressure of the ideological theory that assumes all differences of talent are merely alleged. This is however exactly what one hears out of the mouths of education officials. Indeed many pedagogues still believe this today, sometimes in extreme sincerity. Children going to an "integrated school" come off even worse under the pressure of such an ideology although it was intended to relieve them. What could it mean to someone who as a student, encouraged by his teachers in a thoroughly beneficient manner, tries hard and still has to experience that his classmate succeeds with comparably less effort in areas where he himself fails? The adolescent who has this kind of experience can not learn how to live if the system educationally, politically, and pedagogically denies him the understanding that the process of maturity and becoming an adult means among other things to be able to manage one's limited yet nevertheless unique abilities and opportunities. This has to occur in recognition of the inevitable differences in these abilities and opportunities.

It is totally unnecessary to counter this with the statement that anyone who talks about differences in talent cultivates an ideology which attempts to neutralize socially contingent differences. However small one wants to rate the genetically implied differences of talent – the causality of social, cultural, and in particular family ties to these differences remains a condition which educationally and politically can never be totally erased – not even in a totalitarian social order. So it remains true that attained equality of opportunities results in a differentiation dependent on factors which are principally incompatible for an equal distribution. Nevertheless, the outcome of this differentiation requires cultural and political acceptance – otherwise they will be unbearable for all participants because of their conflict with reality.

In conclusion I would like to explain once again the inconsistency which has resulted in the refusal to recognize differentiated effects in order to attain equality of opportunities, by quoting the former Secretary for Education of the State of Niedersachsen, Peter von Oertzen, who made a great impression on me. During his term in office, von Oertzen negatively commented on certain effects of the newly created promotion of orientation levels between primary and high school. Von Oertzen stated that instead of becoming a valuable opportunity for the encouragement and promotion of young adolescents, this would be scandalously misused as a tool of selection. However, it is exactly this "promotion" as mentioned here by von Oertzen, which makes unalterable natural and social origins visible at the same time. Instead this "illegitimate" outcome of "encouragement and promotion," expressed through the phrase "tool of selection," was the item that should have been declared scandalous.

It requires no further proof that recently the same contradictory denial of the differentiating effects for attaining the equality of opportunities damaged our scientific culture considerably. In extreme cases, out of a supposed "obligation" to offer privileges with the intention of establishing an equilibrium to people of lesser abilities, the result was a kind of shocking practice of giving away discounted grades, where graduates of certain courses of studies at "integrated universities" received inflated grades. The effect was that these students, who had been favored, could immediately continue their studies by entering medical school while neighboring "high school" graduates with their correctly administered yet less favorable high-school diplomas had to be put on waiting lists.

Second, and analogous to this, university research was also to some extent put under the pressure of a supposed obligation to establish a more equal balance through the discounting of excellent achievements. From the nineteenth until the middle of this century, as everybody knows, one of the most effective means for the creation of a research elite within the university was to have an agreement about the appointment between the respective state office and the professor who was about to be called for a university job. It is unnecessary to deny that this old and quite frankly unrevivable system led to an unequal distribution of research means within the university, often dependent on criteria which were only casually related to the purposes of promoting top research. This was especially true in terms of favors concerning appointments of professors which depended on rationed shortages of junior scientists in their respective subjects. On the other hand, there were also disadvantages in this system of appointments even of first-class scientists in some crowded disciplines. Mainly, however, within the scope of the abovementioned practice of appointments, there was a system of unequal distribution of personal and material means which was oriented towards technically proved ranks of achievement.

This system of person related, elite creating research promotion has in the meantime almost totally collapsed. Actually one can not well imagine how a public university administration could be able to negotiate with members of a university faculty of a much larger size concerning appointments or sabbaticals in a manner that would consider all individual competences. The pressing fiscal forces do their own part. In any case one has to realistically assess the ability of academic bodies of administration to unequally distribute their administered resources depending on achievement as being very low. All these reasons result in the fact that the research resources within the university have the tendency to be distributed according to a "watering can principle," which more often distributes less and less.

Top research everywhere requires the efficiency of the so-called Matthäus principle of the social scientist Robert K. Merton: "To him who has, more will be given." The effectiveness of this principle within the universities has weakened. Precisely this matter concerns some of the most significant damage of research and political changes during our university reformation. The question is how can this damage be rectified without, in effect, simply returning to the old system?

Third, it is in connection with the history of our university system (particularly with our system of old state universities), that Germany for the most part developed equally ranked universities. This was done over a long period of time beginning at the start of the nineteenth century, during which German universities underwent various reforms. Although there were certain differences in prestige which distinguished whole universities (as in the case of Berlin) or certain special areas (e.g., the natural sciences in Göttingen), even between Berlin and Königsberg there were never any differentials of academic prestige in Germany like the ones between Yale and Harvard, on the one hand, and colleges in the Midwest, on the other. One could say the same in comparing the German university system with the British. Consistent with traditional French centralization, the special position of Paris always remained unique in the scientific world, even up to today.

If the theory of quantitative expansion leading to a higher differentiation of quality is true, then equal ranking of all institutions of a tertiary scientific educational system, as well as the broad development of this system in the long run is not socially feasible. Naturally, this had been recognized fairly early and plans for the establishment of universities with prestigious special research and political centers are based on this understanding. This has been shown in the founding charters of the universities of Konstanz and Bielefeld, which have in the meantime become recognized historical documents. Certain remainders of peculiar demands which in the first half of the 1960s still could be expressed rather unaffectedly could survive at some places – of course elsewhere as well because of other reasons – until today. Overall however university expansion especially in our country was connected to an ideology which was not favorable for the consideration of special prestigious research and political demands. In other words, it should be possible that the university in Radevormwald could be of the same top quality as the one in the neighboring city of Cologne. Through this, university politics developed into a policy which, after equalizing all in-

stitutions of the secondary school area, proceeded to equalize
the highest, namely, the university level. Looking back later
on at the trend of this attempt at academic equalization on
the highest level, it can be seen that its straightforward interest at the center was to equally promote people of all so-
ciological and economic classes involved in tertiary education into the highest class. Naturally this intention, especially for various unions struggling for a new distribution
of wealth which was being fought by educational and political means, was legitimate. This is not to say, in all due respect
for the demands of this particular social justice, that a change
in the interests of this matter was not socially necessary. In
any case there was causality presented that e.g. primary
school teachers would have to have been educated at universities. Even in other countries, for example, in neighboring Austria, which has been run by a socialistic government
continually for more than a decade, attempts to mask sal-
aried political interests for higher rankings by such pre-
tended causalities remained without success.

This tendency towards institutional university equalization, particularly from a legal aspect as effected in our
country within the past 15 years, could naturally only be followed on the material side in a very inadequate manner. This
was already brought about by the limits of rational fiscal
means. Thus today the university scene from north to south
(for example between Munich and Oldenburg on the one
side and Kassel and Bonn on the other) which although
against the original intention of this "equalization" is nevertheless more colorful and differentiated than any university
scene in Germany since Napoleonic times. It still remains to
be seen whether this is a viable opportunity and not a regrettable deficiency under the aspect of an elite theme.

Fourth, the factual equality principle of status and prestige
of all universities is thus incompatible with the outcomes of
mass academicization. Analogically it is truly fictitious that
all academic education courses are equally ranked throughout the entire Federal Republic and therefore could also
create equal career expectations upon graduation. Nevertheless even today, our public service regulations are a particularly poignant example of this fabrication. A comparison

with those countries where there is neither legally, culturally, nor politically any doubt about the career importance of special elite education, teaches us the incompatibility of this fabrication; a differentiated outcome based on quantitative expansion of our education system can not be suppressed in the long run.

Here I would like to interrupt the enumeration and explanation of the obstructions to this differentiated outcome of a broader development of our university system and, in conclusion, name and explain several possibilities at our disposal which would effect a stronger differentiation of outcomes through realization of the principle of equality in the German scientific system.

1. The cultural acceptance of this differentiated outcome of a realized principle of equality remains by far the most important requirement. Whoever still discredits this differentiated outcome of challenge and achievement has not understood the sociopolitical implications of this process at all. One grasps them if one for example would envisage the outcomes of a distribution of income of mass education. Indeed, the expansion of our student population has resulted in a drastic extension of the circle of those employees who can be happy today about their academic wages which include a splendid starting level of "A 13" in the public services. "That is exactly what we wanted," has been the educational and political standard comment to this fact. But the more one acknowledges this educational and political intention as being fundamentally justified, the more one should also be bound to envisage the reverse of this process. One does this by asking at whose cost this mass redistribution of income took place and for the benefit of which part of the population has it occurred? The answer has to be: It happened to the relative disfavor of the, by far, bigger part of the population which remains without high school or college diploma. One can not reasonably expect that particularly the education and science unions could consider this to be a problem. But one may be anxious to know the point in time at which other trade unions, being different from the education and science unions, will have recognized that the outlined results of a redistribution of income within mass education do not cor-

respond to the interests of their members. It is foreseeable that the aforementioned – by far greater – nonacademic part of the population in the long run will only accept the mentioned financial/political outcome of education (which especially in Germany is linked to public services) if the connection between effort and achievement in a study program on one hand and a prestigious business position acquired later on the other remains visible. That means that the short-lived epoch is quickly coming to an end where it was sociologically possible that one could reach otherwise unattainable career opportunities through programs and examinations which were weakened by a reduction in academic standards. Meanwhile common sense has dictated a new understanding that social conditions which open up for somebody based on his academic excellence must be tied to reliable verification of qualifications if they should be taken as justified. To repeat, in the end, there can only be an elite in a public culture that acknowledges an elite class and at the same time adheres to the particular requirements for achievement by which an elite would have to be measured.

2. An acceptance of the development of differences in prestige between various institutions within the German university system and at the same time the promotion of this development is long overdue. Principally it can not be opposed that the foundation of private universities which indeed are not known in the traditional German university system can also contribute to this overdue, "elite-promoting" differentiation of our university system. Nevertheless, one has to add that the famous American model produces merely an illusory effect. First of all, one has to realize that the relative importance of private universities in the USA has not increased but decreased due to the pressure of the escalating costs of the American science system. In addition (not to mention the special German legal conditions for their foundation) the number of private universities which at best could be placed among the public universities in Germany would always remain too small to be able to trigger an extraordinary competitive effect. For that we lack potential private sponsors and institutions of the needed size.

To repeat, there is no principal objection to private universities. But as an instrument for the promotion of an elite this would be neither sufficient nor effective in itself. It would be comparably more effective to encourage the "promotion capability" of the existing university system. For this purpose it would be necessary to encourage the means necessary to promote personal mobility in the university system, especially of those who would bring better people to better positions. This strategy includes many substantial items, yet even the little things can contribute to its success – such as restoration of the old flexible salary system for university professors and thereby promoting scientists of extraordinary quality; granting additional funds based strictly on a rigid graduated pay scale would be thoroughly counter-productive here. Additionally, here it is also true that the material aspect is incomparably less important than the aspect of due recognition. To exemplify, once again, this connection to the idea of private universities brings to mind the fact that the particular political party which currently favors this idea, in the late 1960s and early 1970s under the banner of the fight against the so-called common university, in a special way, contributed most to destroying the atmosphere of recognized excellence without which an elite can not thrive in any institution. An elite, as has been correctly said, can not be purchased – not even through the implementation of private universities with lots of money. We will only have a research elite if we also publicly accept the consequences of differentiated results based on an equality of opportunities in the science arena as being both inevitable and necessary. In addition we can have a research elite only if we continually pursue the promotion of an elite class, the likes of which has already been established, instead of placing the elite without hesitation in the winds of culture-revolutionary ideas at that point in time when the so-called integrity of a political party, which normally likes to keep their extremistic wings contracted, apparently demands it.

3. The promotion of a differentiated university system by means of new types of elite-favoring institutions is more important than the mobilization of new private sponsors and institutions. For example, one may think here of the "Histori-

cal Kolleg" in Munich. Although small and relatively young, this institution has proven itself as an extremely renowned society for the promotion of research excellence. Through a visiting fellowship at the "Kolleg", researchers are provided a reprieve from their ordinary academic and professional duties in order to finish more prominent, well-prepared projects.

I repeat that this exemplary "Historical Kolleg" in Munich is a very small institution. But the purpose of this institution shows the needed direction: special promotion of values which have already been proven as being ideal.

One should remember that the aforementioned "Historical Kolleg" is the result of a comprehensive idea of previous euphoric years, namely, the idea of a "German College of Science." It is superfluous here to name the reasons which allowed this. However, the idea was too much oriented towards the almost unimitable French model, which remains unattainable up to now. Nevertheless, there is nothing against the idea of founding similar institutions at other locations for other subjects analogous to the "Historical Kolleg" in Munich. To be sure, for quite some time there has been substantial financial support offered for such plans.

Obviously esteemed reliable academic scholarships, as granted by such foundations as the "Volkswagenwerk," also belong in the general design. Special research areas of the German Research Society have been particularly effective in promoting exceptional results.

It is superfluous to describe this process in detail. Only a few things have to be newly created here. The rest would occur by way of a snowballing effect, feeding upon its own success. The basic principle in all of these cases is clear: recognition of differentiated outcomes through the expansion of our science system by means of active promotion of important scientific innovations.

4. The "Historical Kolleg" here mentioned as an example owes its existence to the well-known initiative of its founders. This makes it clear why the function of large science foundations did not by chance decrease but instead steadily increased with the expansion of our scientific system. Institution foundings effectively contribute to the building of an

elite through scientific promotion according to a type of "top dressing" principle. Those areas in the research arena which have already developed in a very promising manner, receive what they need to develop even more than originally expected. This is the sense of the well-known "Matthäus principle" without whose validity you can not have any achievement in science.

I repeat, particularly in the framework of the current German science system, the importance of the described role of private foundations does not decrease; on the contrary it increases. It is most regrettable that the tendency towards such private foundations for the fulfilment of this function has recently not increased but rather decreased. Why is this the case? By far the most important reason for this regrettable tendency is the following. While the public science budget is essentially able to follow the increase of personal and material costs in research and to compensate inflation losses, there is no equivalent expansion of foundation capital whose returns should favor additional research. That means that the "relative" share of private foundations, looking at overall research expenditures, continually diminishes. Here it is unnecessary to reflect the drama of this process by utilizing statistics. The result of these numbers would say: private foundations are slowly becoming an "endangered species."

Principally there seem to be only two ways to counter this danger. One way would be to use a part of the returns of foundation capital for the purpose of a research capital increase. However this way is presently not feasible since, according to the federal finance department, this can not be recognized as a nonprofit use of foundation capital.

This type of classification can not be easily understood. It would be understandable if it would affect fiscal interests, i.e., if the partial use of the increase of foundation capital would mean some costs to the state. But it requires no costs to the state. The capital returns, necessary in a nonprofit sense for an increase in capital, to secure the ability of these foundations to contribute to the extent they do now to research promotion in the future, are neutral one way or the other in the sense of public finances. By not recognizing this kind of use of capital returns through private foundations as being a public "utility," the relative contribution of such foundations to research in Germany will be continually reduced. This is nothing less than a mechanism administered by the financial authorities which, in the long run, will totally eliminate such foundations as relevant factors in the promotion of research. If this mechanism will remain in effect, what is the benefit of such legislation?

The other additional course would be to eliminate some of the legal impediments which today inhibit the establishment of private foundations for the purpose of nonprofit research promotion. The corporation tax law of 1977 which in addition heavily affected existing foundations, is one of these impediments. Indeed, the revision of this law would be costly. To expect the state to incur these costs even in the present situation is not necessarily unreasonable. The institution of the corporate tax law of 1977 was rather unreasonable with its drastic deterioration of the legal tax conditions which have to be observed by private foundations. Even today, no significant common good can be recognized as a side effect which could justify this deterioration. Accordingly it is fair to expect that private foundations might finally be put again under their former conditions. Only this could revive latent willingness to develop new foundations.

Could there be any powers in our country who do not want this? There are some at the edges of the spectrum. Predominantly however the science foundations enjoy significant political recognition explained by their highly necessary function. In the state address of the chancellor in May 1983 it was announced that the government "will promote the existing foundations and check how new, nonprofit foundations could be encouraged." All groups of parliament at that time already in 1982 spoke up for a political correction of the private foundation law in relation to the damage caused by the corporation tax reform. Out of both motives taken together – out of the objective pressure which the science foundations are subject to today on the one hand and out of the unanimous scientific and political recognition of the necessity and importance of their work on the other – it should be possible to devise a favorable prognosis for the rapid improvement of conditions of foundation activities. All participants and those affected certainly hope for it.

5. There have also been certain negative effects of ideological superstitions, which hold that proximity to industry and economics would endanger the higher ideological purposes of research. This has been in addition to the damage caused by reforms to which the German university system was exposed from time to time in its rapid development in the late 1960s and early 1970s. In the meantime the exact reverse has been discussed as true. The concentration merely on ideas of economic usefulness only then becomes a misleading orientation if it endangers the logical preference of basic research. Naturally, only applied research tied to basic research will remain useful in the long run. Under these assumptions a developed pragmatic sense of usefulness and the ability for an efficient application of research results are very important criteria for research elites themselves upon which we are dependent today. In connection with this, one has to wish that the legal obstructions, in reference to civil servants and secondary employment, of a liberal transition between practical research and entrepreneural business will be quickly reduced. The understanding of the imperative nature of this is steadily increasing everywhere in the public administration.

One could go on for quite some time concerning the imperative nature of this recognition and confirmation of a differentiating effect in promoting an elite through the quantitative development of the German university system.

A scientific elite in the described sense has neither rank nor class attributes. An elite represents differences which will become visible exactly at that time when ranks and classes have disappeared. It is envy that denies its recognition. The masses are not favored by this envy but at best only those with ambitions for establishing themselves as the "functionary" elite by the political use of the emotional potential of this envy intentionally fanned.

Future Information Technology –
Motor of the "Information Society"

GERHARD KRÜGER

The terms information society and information technology have recently, with impressive speed, become slogans in our everyday discussion. If so far mainly small circles of experts from the fields of science, economics, and politics, or from the trade unions have occupied themselves with the growing possibilities of the new information technology and its potential economic and social effects, it can be stated that nowadays the mass media – themselves fundamentally concerned by such innovations – make themselves heard with contributions pertaining to these developments and the consequent and very differently colored pictures of the future.

At present, the mostly unprepared and thus often helplessly reacting public is being flooded with radio and television programs, with special books bearing more or less exaggerated titles, with newspaper articles and essays in magazines – all concerning the new technologies.

Social critics of all persuasions have seized upon the topic. Orwellian visions are being conjured up, with the robot as job killer, the computer as policeman of the supervized state, cable television as a drug and ruling instrument, the future information society a nightmare bringing the end to civilized human society. A halt, a change of views, a renunciation of "Faust's pact" are being called for. Technology moratoriums or a stop to all developments in information technology, enforced by violence if necessary, are being demanded by some.

On the other hand, there are the representatives of a forward-looking strategy used for coping with the undoubtedly

far-reaching changes that society will be confronted with in
the wake of information technology.

They proceed on the assumption that the way towards an
information society is already paved for in the highly industrialized states of the world, that the Federal Republic of
Germany with less than 1.5% of the total world population
will not be able to exercise a guiding influence on the development of world technology, and that our country will not
succeed in holding its position as a leading exporter of industrial products without a strong position in the high technologies. A stepping back by the Federal Republic of Germany into the rank and file of the industrial countries,
weakening of its international competitiveness due to an atti-
tude of refusing technological innovation, from this point of
view, entails by far more aggravating negative consequences
for the labor market and society than would an active tech-
nologically innovative configuration of the impending structural changes. It should also be mentioned that the new
technical possibilities do not merely present risks, but, when
applied responsibly, offer good prospects for fulfilling the social objectives of a future postindustrial society.

At this level of discussion, which obviously is not always
held emotionlessly, it will surely be helpful to cast a short
glance at the history of our key term: information society.
The first observation of this retrospective, which might be
surprising to some people, is that the term information society has been in international usage for hardly more than 10
years. To the best of this author's knowledge it appears for
the first time in a study entitled: "Information Society of the
Year 2000," made available by a Japanese group of experts
from industry, economics research institutes, universities,
and the media and carried out by order of the government.
This study was published in 1972. Remarkably, it just so happened that exactly in that year the microprocessor in its most
primitive form was developed by a small American firm – for
the first time – incidentally, consisting of a considerable
number of pieces. The following should once again be
brought home to all those who have reservations towards the
present technological forecast concerning the worldwide
speed of development in information technology: it has only

been 12 years since the central product of microelectronics, namely, the microprocessor or, in its further development, the microcomputer has come into existence. Look how powerful its determinative force is already for the economy, world trade, and changes in international competition!

The vision of the Japanese in 1972 was courageous and of outstanding clarity. Although there was no real experience with the widespread use of highly integrated microelectronics (the first microprocessor had less than 3000 switch elements on one chip, today for the same price and on a comparable chip size there are 100000 and more elements!), the way towards the new information technology which was determined by microelectronics was perceived very clearly and was predicted, involving the integration of computer technologies, telecommunications, consumer electronics, watch and camera technics, as well as industrial automation (particularly robot technology).

Even the far-reaching economic and social transformations which were to be expected as a result of the new basic innovations were at that time already specified by the Japanese with astonishing accuracy.

Government and industry in Japan did not merely prophesy these assertions, which today are self-evident to experts the world over, but which were revolutionary at the time. The suggestions of the experts did not disappear in counselling commissions, rounds of hearings, and finally the files.

The Japanese combined their energy into national policy with the aim of creating, within a few decades, the economic prerequisites necessary for the structural changes forecast without causing any damage to the Japanese economy and society; this was to be achieved by an extension of their semiconductor industry, their computer hardware and software, and the applications of information technical components in many products. As a footnote, it should be said at this point that during the realization of this program in which the Japanese are still today very engaged, they have managed to get by without any huge government subsidies except in the fields of basic research and basic industrial development.

The enormous research and development expenditures of the Japanese information technology industry have been in-

directly assumed to an essential extent by their main com-
petitors, for example, the Federal Republic of Germany, by
imports (worth billions) of Japanese consumer electronics
and optics, including watches, cameras, pocket calculators,
video recorders, video cameras, etc.

In review it might be inexplicable, even a mystery, why
these Japanese considerations which were made known at an
early stage attracted attention in the United States but not in
the Federal Republic of Germany. Maybe the missing re-
action gives some evidence of the dynamics of West German
society, in which context the scientists themselves should not
be exempt of criticism, as they obviously failed to warn em-
phatically enough.

It was only the report "The Informatization of Society,"
written in 1977 by order of President Giscard d'Estaing of
France, and well known by the names of its authors, Simon
Nora (who was general inspector of finance) and his collabo-
rator Alain Minc, that drove home to those in Europe the
idea that today's world driven by the basic innovations in
computer technology, microelectronics, and informatics is
nearing a new – the second – industrial revolution. By the
way, it is also characteristic of the speed of development and
the difficulties of technological forecasting that the Nora
Minc-report is already obsolete with regard to its essential
technological prognoses; this fact does not, however, detract
from its structural and sociopolitical conclusions.

While Europeans were writing studies and while govern-
ments in England and France were directing flaming appeals
towards science and industry to participate in the innovation
race, the next decisive step – which was not previewed by any
planning bureaucracy – was taken in a garage in Silicon Valley, California. Two American computer engineers, both very
young, hardly more than 20 years of age, built the first personal or home computers. These game and home computers
and work stations, which meanwhile run to millions of production units per year, have almost become the symbol of the
new technology for the everyday man.

From the economic point of view, it is to be stated once
more that, in the case of the individual computers, as with
consumer electronics and video technics, another market

worth billions of dollars in information technologies passed us by, with the one exception that this time the Americans came in first.

The Federal Republic of Germany was once the "Purveyor to Her Majesty" of the world of classical information technology, from optics and fine works technology via telecommunications up to wireless engineering. How in the second half of the 1970s until the beginning of the 1980s Germany has taken up the worldwide challenge in information technology can be illustrated by a single figure (perhaps exaggerated, but striking): Between 1974 and 1982 the federal government reduced its advancement subsidies for data processing, informatics, automation technology, and microelectronics nominally in German mark amounts by more than 50%; really of course (due to inflation), even more drastically. The "Bundesländer" (the states of the Federal Republic of Germany) also made their antiinnovative contributions for mastering the future. Thus, the further extension of the programs of "informatics studies" (computer science and engineering) which, as of 1970, had been set up in approximately 14 universities was stopped in all Bundesländer effective 1977. By freezing capacity, informatics got away relatively favorably in the southern Bundesländer. But further north, in the more precipitous sections of the south-north incline, the number of study places for informatics was once more clearly reduced by curtailments of staff and financial support.

Sincerity calls for adding mention that the reduction measures for informatics in fact was very moderate in Baden-Württemberg. As the state government of Baden-Württemberg actively pursues research and technology politics, an evident return demonstrated by a resumption of the extension of informatics and other fields in information technology has recently come about. The program of the state government also provides for improved computer equipment for all scientific-technical branches at universities and academies. Both measures are of eminent importance for the overall economic development in Baden-Württemberg and are greatly appreciated by scientists.

But if we sum up the present international situation, we must state that the superpowers in informatics and microelectronics, the United States of America and Japan, with an enormous amount of resources are engaged in a neck-and-neck race for the top future positions of the world information technology industry, meaning the overall world economy. In the lee of the great powers, several East Asian states, like Taiwan, Singapore, Hong Kong, and South Korea, are securing considerable shares in the international market of top information technology products, a fact which forces especially the Japanese to constantly turn in peak performances in technological innovation. In Europe, England and France try to stay in the front row by means of intensive technological advancement and also by trade measures, whereas the Federal Republic of Germany unfortunately has to register some failures and misfires in its information technology motor – which had been so efficient in the past. After all, it should alarm every responsible person that, according to figures of the Statistical Federal Office, the Federal Republic of Germany, both from the standpoint of technology as well as application, since 1978, without an interruption, has had a negative balance of trade in the key field of office automation and information technology. Even when one includes the domestic production of the subsidiaries of the computer manufacturers from abroad, the volume of domestic production of office automation and computer technology is still much smaller than the domestic requirements (a situation which is very well taken advantage of by worldwide competition, as can be readily observed, for instance, by taking a look at the computer shops or the computer departments in the big stores).

In the first part of this lecture, I deliberately dedicated some time to tracing the history of the development of modern information technology which is based on microelectronics and to showing the differing willingness of economic and political groups in the various parts of the world to accept the challenge of the transition to an information society. We require this view on the starting position if we want to gain a profound comprehension of future developments.

Our look at the history of the new technologies – although strictly speaking it is daring to talk of "history" in the case of a 12-year-period of information technology in modern times (which is what has elapsed since the appearance of the microprocessor) – shows, in the first place, the extraordinary speed in carrying through every one of the information technology innovations to the level of mass usage and even up to worldwide mass markets.

The extraordinary enactment speed is valid both for the markets of private-end consumers of, for example, video recorders and home computers, and for capital goods markets of, for instance, robot technology, computer-aided design and manufacturing, or the new digital telecommunications technology.

Reliable forecasts proceed on the assumption that only 10%–20% of the (in this century) efficient utilization possibilities of microelectronics and informatics have been so far started upon at all. In the coming 16 years (until the turn of the century), we could thus expect five to six times the technological innovation potential since point zero of the microprocessor. From the history of technology we know quite well these trend curves – considered only qualitatively – to have an initial period which is not very striking with regard to the whole national economy, followed by a rapid penetration phase which is of prime importance to the economy and society. Take, for example, the expansion of the utilization of automobiles in the Federal Republic of Germany within the past 30 years, although the history goes back almost 100 years.

In using the example of the telecommunications services offered by the "Deutsche Bundespost" (the German PTT), one can – for a partial survey of the future field of integrated information technology – already today outline the type and volume of these changes within the next few decades quite well.

The fundamental technical principles of the telephone have essentially remained the same within the 100 years of its existence. Its most important improvements for the subscriber during this century have been the long-distance call even beyond continents, and the automatic dialing system. Gradually over a period of decades their utilization has become generally widespread – at a rate determined by economic rather than technical conditions. But, as a general principle, the kinds of utilization and the operation of the telephone sets have basically remained unchanged for the past 60 to 70 years.

Starting in 1985, with the transition to a digital telephone network, a field in which German electrical communications engineering and the "Bundespost" lead the world, the technologies and the connected utilization possibilities of the ancient "audiofrequency telephone" will be revolutionized. Within the extended offer of the "Bundespost," which results from the digital exchange techniques, among other things we find "call transmission;" at the request of the subscriber the postal computer can, within a certain time limit, transmit an incoming call to another call number. We also find the "baby call," in which case a call with a previously determined telephone connection is put through by pressing any key on the key telephone. The service "Do Not Disturb" which is available for the subscriber on account will surely enjoy great popularity; the PTT computer intercepts incoming calls and informs the caller that the subscriber does not want to be disturbed at this moment.

Much more important than these luxury improvements in telephone services, however, are the changes resulting from digital communications engineering. The basic technical principle of transmission has changed to such a degree that all the possible ways of information presentation, like for instance picture, text, and language can be transmitted – mixed jointly and arbitrarily – between the owners of modern digital telephones. The present "telephone," which is of course only suitable for the transmission of spoken language, will be replaced or extended by more universal telecommunications terminals. This development has already been put into action by means of the interactive videotex. For the acceptance of the new digital communication engineering, it is surely a vital point that the utilization of the new communications technology takes place by using the already existing telephone cables. Thus, for setting up the digital network it is not necessary to tear up streets, dig up front gardens, or knock down walls.

If one completes this picture with the digital mobile telephone (known today as the auto telephone) which can be designed more compactly and more cheaply by means of computer technology, then it becomes obvious what a far-reaching radical change, visible for everybody, will be taking place in the classical communications technology of the telephone – starting already in this decade.

As a second example for the powerful dynamics of the new technologies, let us consider the so-called personal computer. Up to now, no really appropriate expression has existed for this computer and communication technology device which is meant for individual use; the present names like personal computer (PC), home computer, or work station are rather more confusing than elucidating, especially with regard to the future role which it will play for the individual.

In what follows, the key functions of the individual computer, as foreseen for the 1990s, will be roughly outlined against a background of coming social changes.

Basically, the experts are in agreement that, in the future society of the industrial countries, the following facts will be of significant weight: In professional life, education, and leisure time, in the course of satisfying basic needs, organization of one's personal environment, and especially in the social and political field, dealing with large quantities of information will play an extremely important part for the individual, the social group, and society as a whole. One should not, however, jump to the faulty conclusion that the possibilities of information technology providing rapid and cheap procuring, distribution and filing of information are in themselves primarily responsible for giving rise to the snowballing expansion of information, namely, the information inundation in modern society, the control of which becomes more and more problematic at all levels of the human community. Our highly complex and meanwhile planetarily expanded total economy, with the worldwide social interdependencies attaching thereto, can only be balanced by an increase and differentiation of information. For instance, there used to be many regional markets which were interdependent to a very small degree only, whereas today we find a uniform international market, a planetary total economy even up to a very precise division of labor in the production of technical goods. In this manner, technological innovations and production progress in Eastern Asia can, within a period of a few years, wipe out traditional European leaders in the market, as has happened in the case of the Swiss clocks and watches industry, where almost 75% of the original work places have been lost. Similarly complex problems also appear in our individual states; one only has to take a look at the present fiscal and social insurance laws. Information demand and information volume will continue to zoom up, a fact which is enforced by the social and economic problems in and among the countries of a more and more densely populated world, the economic and ecologic crises of which can only be solved by more and better information.

Whatever applies to society, also applies to each individual in a more and more sophisticated information environment: In the future, only those who have sufficient information and technical means at their own disposal, and under their complete control, will be able to survive and participate in an information society. Both, from the technical point of view and from the standpoint of social responsibility, this claim makes the personal computer, or individual computer an indispensable instrument in a future human society which supports the self-determination, but also the professional efficiency of each citizen.

Some functions of the private computer of the 1990s will illustrate the above remarks. To some people, the description may sound very futuristic. Further down the line, however, it is a question of innovations, the principal technical realization of which – though not necessarily their profitability – are to be regarded as assured, even today.

As an essential partial function the personal computer handles its different communications tasks on the same basis as the digital telephone connection, which has already been described. Thus, it will also undertake the tasks of the present telephone and the ones of an "intelligent" answering machine, which – in an advanced form and according to the instructions of the owner – will be able to give precise answers to questions which have been posed telephonically or tele-graphically. In its capacity as an electronic letter service, the personal computer will also receive texts and drawings. In the year 2000, the home delivery of paper letters will surely not take place daily any more and the hobby collectors will start preserving and taking care of the PTT letter boxes in the same way that steam locomotives are treated today.

As an image and text editing system, the individual computer is helpful in designing one's own electronic mail and in sending it to the receiving computer via the public digital telecommunications system.

An "intelligent" computer goes even further. Using text analysis and compression procedures (the fundamental aspects of which are known today already and as they are used by computer-aided libraries and data bases), it sorts out incoming electronic mail, selects uninteresting consignments, and compresses long-winded messages to their essence. (On request, of course, it also makes available from its records the original version and produces it on the screen or as a paper copy.)

In the same way as in the case of the chess computer, where the degree of difficulty and the playing skills of the computer are adjustable, the owner of an intelligent computer will also be in a position to adjust its sphere of interest, and its threshold of tolerance for the cognizance of incoming information. Whatever "is not worthy of the paper," the computer, after evaluating the text conveys to the electronic wastepaper basket – without any environmental burden. Important material goes into the electronic files, e.g., the data base of the personal computer.

With the help of the computer, its owner can of course actively request information from data bases, place orders, make banking entries, etc. The individual computer with a telecommunications connection is thus able to perform all the services which are in the process of being introduced under the name of "Bildschirmtext" (interactive videotext).

The second main function of the private computer is to support the owner in all aspects of personal information processing and storage. In its electronic archives go personal records, daily memorandums, one's reference library, as well as all sorts of address lists, notebooks, and memobooks even up to fully up-to-date maps which are stored electronically or optically and with the help of which the owner can, for example, plan a trip according to the latest conditions (including construction on the motorway). But the personal computer is not only a storeroom for notes, it is indeed a genuine personal knowledge base.

This base contains adopted knowledge, meaning acquired and learned knowledge, but also – and this is of primary importance – individual knowledge, meaning knowledge which was actively found out by each single individual.

In informatics, we would say that the personal computer contains the objectified portrait of the knowledge, capabilities, and very personal experience of its owner in an automated operational form – and available only to him. It is a system of knowledge of a very particular individual, a collection of his/her intellectual property.

These prospects render the personal computer, and the accompanying digital communications infrastructure that will ensue, the decisive contribution among the future information technologies to the formation of the information society of the next century. It will not be the exclusive application of large computers in protected computing centers by a small group of specialists, a power elite of computer scientists and politicians, which will form this society, but, on the contrary, it will be the individual computer at each place of work and in every private home.

In the case of simple and average information and communications requirements, the operation of this computer will then be extremely unsophisticated and uncomplicated. We already get an indication of the very intelligible forms of interaction, as they can be observed very well in the case of demanding computer games or in the case of the most modern small computers with color graphics and the so-called "mouse."

It is indeed this recent development which gives rise to the assumption that, after bringing to perfection the respective software-supported man–machine interaction for personal computers, there will probably be less computer illiterates than reading and writing illiterates.

Colored and lively computer information presentations can be in fact much more vivid than a printed or written "static" text. A far-off future generation could possibly consider a book in its present form, a typical example of the least philanthropical method of information presentation.

The theses of this contribution can be summarized as follows. An increasing number of people are starting to realize that the world is in the process of undergoing a dramatic economic and social change, the direction and speed of which is primarily determined by the highly industrialized countries. The past conceptions of order and value of the nineteenth century which were very strongly influenced by the social changes resulting from the first industrial revolution, have to make way for the new structures of the information society. The transition to an information society – like all the other fundamental processes in history – takes place inevitably and cannot be stopped by any single nation.

The new information technologies are only one of the reasons for the social change. Their rapid development and the resulting intensified struggle for the international market of highly technological goods, however, make the future information technology, consisting of microelectronics, informatics, and telecommunications the motor in the transition process.

In the transition period, the new technological possibilities and the changing economic and social structures will undoubtedly involve a lot of problems and a certain amount of hardship.

But without the full utilization of the future information technology there will not be any solutions, which will be accepted by the majority of citizens, to the present and coming social problems.

To get off the train which is on its way in full swing towards an information society is not conceivable. Attempting to do so would mean a loss to every industrial country of a large part of its economic and political power – with incalculable consequences for international competitiveness, social security, and internal peace. The Federal Republic of Germany which, on the grounds of its economic and social structure, is particularly challenged by the modern information technology, has no time to spare in accepting this challenge by marshalling its economic, scientific, and state resources.

Authors

RUDOLF HENN
Dean of the Faculty of Economics
University of Karlsruhe

LOTHAR SPÄTH
Minister President of Baden-Württemberg

HERMANN LÜBBE
University of Zürich

GERHARD KRÜGER
Institute for Informatics III
University of Karlsruhe